ULTIMATE SUPERCARS

LYKAN HYPERSPORT

By Joanne Mattern

Kaleidoscope
Minneapolis, MN

The Quest for Discovery Never Ends

This edition first published in 2023 by Kaleidoscope Publishing, Inc.

No part of this publication may be reproduced in whole or in part without written permission of the publisher.

For information regarding permission, write to
Kaleidoscope Publishing, Inc.
6012 Blue Circle Drive
Minnetonka, MN 55343

Library of Congress Control Number
2022938002

ISBN
978-1-64519-611-2 (library bound)
978-1-64519-681-5 (ebook)

Text copyright © 2023 by Kaleidoscope Publishing, Inc. All-Star Sports, Bigfoot Books, and associated logos are trademarks and/or registered trademarks of Kaleidoscope Publishing, Inc.

Printed in the United States of America.

FIND ME IF YOU CAN!

Bigfoot lurks within one of the images in this book. It's up to you to find him!

TABLE OF CONTENTS

Chapter 1: Something to See **4**

Chapter 2: New Kid on the Block **12**

Chapter 3: Inside the Car **18**

Chapter 4: Fast and Furious **24**

Beyond the Book...*28*
Research Ninja..*29*
Further Resources..*30*
Glossary..*31*
Index...*32*
Photo Credits...*32*
About the Author..*32*

Chapter 1
Something to See

Ahmed stopped in his tracks. He could not believe his eyes. "Papa, look! Look at that car!" he shouted.

Ahmed's father smiled. "Now that is something to see. Let's take a closer look."

Beautiful and expensive cars were a common sight in Ahmed's home city of Dubai. But this car was the most amazing thing he'd ever seen.

Just then, the car owner came over to Ahmed and his father. "I see you like my car," he said with a smile. "It's a Lykan HyperSport."

"It's beautiful," Ahmed said. "I'll bet it can go really fast!"

"Yes, its top speed is 245 miles per hour (395 km/h). But I save that speed for the racetrack. There's too much traffic in the city to go that fast. I have gone from zero to 60 miles per hour (97 km/h) in just under three seconds."

"Wow!" Ahmed said. "I can tell there is a lot of power under the hood. What kind of engine does it have?"

"This car has a **turbocharged** V6 engine," the owner replied. "It puts out 750 **horsepower**."

"I like how the car slopes down in the back," Ahmed's father said. "That makes the car more **aerodynamic**, right?"

"Yes, it does," the owner replied. "The **carbon-fiber** body also makes the car light and fast. This car has both power and beauty."

FUN FACT
There are only seven Lykan HyperSports in the world.

FUN FACT
The HyperSport was in the movie *Furious 7*.

Ahmed looked closely at the headlights. "Papa, look at how much they sparkle."

The owner smiled. "The headlights are one of the special features of this car. They are made with 420 diamonds on **titanium** blades."

"That's amazing!" Ahmed said. "Someday I want to own a Lykan HyperSport."

"I can understand why," his father said. Then he turned to the owner. "Thank you for talking to us. You are a lucky man to own such a fantastic car."

"It was a pleasure." The owner looked right at Ahmed. "Young man, I hope your dream comes true."

PARTS OF A LYKAN HYPERSPORT

spoiler

six-piston brake calipers

CRAZY DOORS

Some luxury sports cars have gull-wing doors. They open out and up like a bird's wings. However, this design takes up a lot of space when a person opens the door. The HyperSport has a better idea. Its doors slide up and back. This takes up less room. It also looks really cool!

reverse door system

carbon-fiber body

diamond and titanium headlights

Chapter 2
New Kid on the Block

The Lykan HyperSport is made by W Motors. W Motors is new on the supercar scene. It started in 2012. At first, W Motors was located in Lebanon. In 2013, it moved its headquarters to the United Arab Emirates.

Lykan was founded by a young businessman and car designer named Ralph Debbas. Debbas named the car after the Lycans. The Lycans are also called werewolves. Werewolves can change from ordinary animals into creatures filled with power. Debbas had the same idea for his new car. It looked beautiful. But step on the gas, and the car turned into a ferocious beast.

FUN FACT
W Motors was the first company to make high-performance luxury sports cars in the Middle East.

Supercar fans got their first look at the HyperSport at the Qatar Motor Show in 2013. The car shown there was just a **concept**. It was not the same as the cars that would be made later.

Many people had heard about the car before the show. Some of them could not believe what they heard was true. When they saw the car at the show, they got very excited. W Motors had really created a car like no other in the world.

ATTENTION, PLEASE!

Ralph Debbas started designing the HyperSport in 2006. That was before he founded W Motors. Debbas did not want to make an ordinary car. He knew that people would pay a lot of attention to an expensive car. So he included many top-of-the-line features. That made the HyperSport stand out.

In 2014, a road-ready model of the HyperSport was shown in Monaco. The car officially went on sale in November 2014.

The HyperSport was a beautiful and powerful car. But it got attention for other reasons too. The biggest shock was its price. The car cost 3.4 million dollars. It was one of the most expensive cars in the world.

Another surprise about the HyperSport was how rare it was. Lykan only made seven HyperSports. In spite of the high price, the company had no trouble selling all of the cars. Many people wanted a chance to own one.

WHERE WAS THE LYKAN HYPERSPORT MADE?

Torino, Italy

Italy

Dubai, U.A.E.

United Arab Emirates

MADE IN ITALY

W Motors's headquarters is in Dubai, United Arab Emirates. But the Lykan HyperSport was made in Torino, Italy. It was made by a company called Magna Steyr.

Chapter 3
Inside the Car

The Lykan HyperSport looks amazing on the outside. The inside of the car is very special as well.

FUN FACT
Unlike some other supercars, the HyperSport has rear-wheel drive.

The car's seats are made of leather stitched with gold. But that is not the most unusual thing about the inside of this car. One of the car's most amazing features is its display. That display is a **hologram**.

When the car is turned off, the display is invisible. It just looks like a blank box on the dashboard. Turn on the engine, and things get exciting.

The empty box lights up. It becomes interactive. Drivers can touch the screen. They can even push their fingers right into the display. Or they can turn on the controls with a wave of their hands. It's an unusual way to turn on music, set the temperature, and control other parts of the car.

Luxury cars should sparkle. The HyperSport really does. Its LED headlights are made of white gold and titanium. Owners can choose 420 precious stones to add to the headlights. They can choose diamonds, emeralds, or sapphires.

POLICE POWER

Police cars need to be fast. The police department in Abu Dhabi, the capital of the United Arab Emirates, might have the fastest car of all. In 2019, Abu Dhabi police added a HyperSport to their fleet of cars.

THE LYKAN HYPERSPORT
IN DETAIL

Height: 3.9 feet (1.2 m)

Width: 6.6 feet (1.9 m)

LENGTH: 14.8 feet (4.5 m)

WEIGHT: 3,042 pounds (1,380 kg)

TOP SPEED: 245 mph (395 kph)

TIME FROM 0 to 60 mph: 2.9 seconds

The HyperSport has a V6 engine. That is small for a supercar. However, that little engine packs a lot of power. The 3.7-liter supercharged engine creates 750 horsepower. The engine is made by Porsche. It roars with power. A dual-exhaust system rumbles when you step on the gas.

This car has a six-speed manual **transmission**. Step on the gas, shift into gear, and let the HyperSport take off!

Chapter 4
Fast and Furious

Over the years, the *Fast and Furious* movies have shown off plenty of amazing cars and crazy driving. Producers knew they needed something special for the seventh movie in the series. So they turned to W Motors and its Lykan HyperSport.

Furious 7 was filmed in 2013. It came out in 2015. In one of its most exciting scenes, characters Dom Toretto and Brian O'Connor, played by Vin Diesel and Paul Walker, need to steal a computer chip from the dashboard of a Lykan HyperSport.

The two men find the bright-red car locked inside a vault. The vault is in the Etihad Towers, a skyscraper with five towers in Abu Dhabi.

FAKING IT

Of course, Vin Diesel did not really drive a HyperSport between two skyscrapers. The scene was created through movie magic. The car traveled on a track and burst through breakaway glass. Then it dropped into a big container of boxes a few feet below. In other scenes, stunt drivers handled the car.

The two men are surprised that someone would keep such a powerful car in a vault. Dom Toretto says, "There's nothing sadder than locking a beast in a cage."

A few minutes later, Dom and Brian are in the HyperSport. They're racing up and across the towers. The out-of-control car roars and spins through an art gallery. Brian O'Connor rips the computer chip from the dash. The two men dive out of the car just before

it flies through a window and crashes to the ground below. It is one of the most breathtaking scenes in any of the *Fast and Furious* movies.

Furious 7 shows off the HyperSport's raw power. The engine roars. The car travels at incredible speed. It's very unlikely that a car could jump between two skyscrapers. But the power of the HyperSport makes it seem possible.

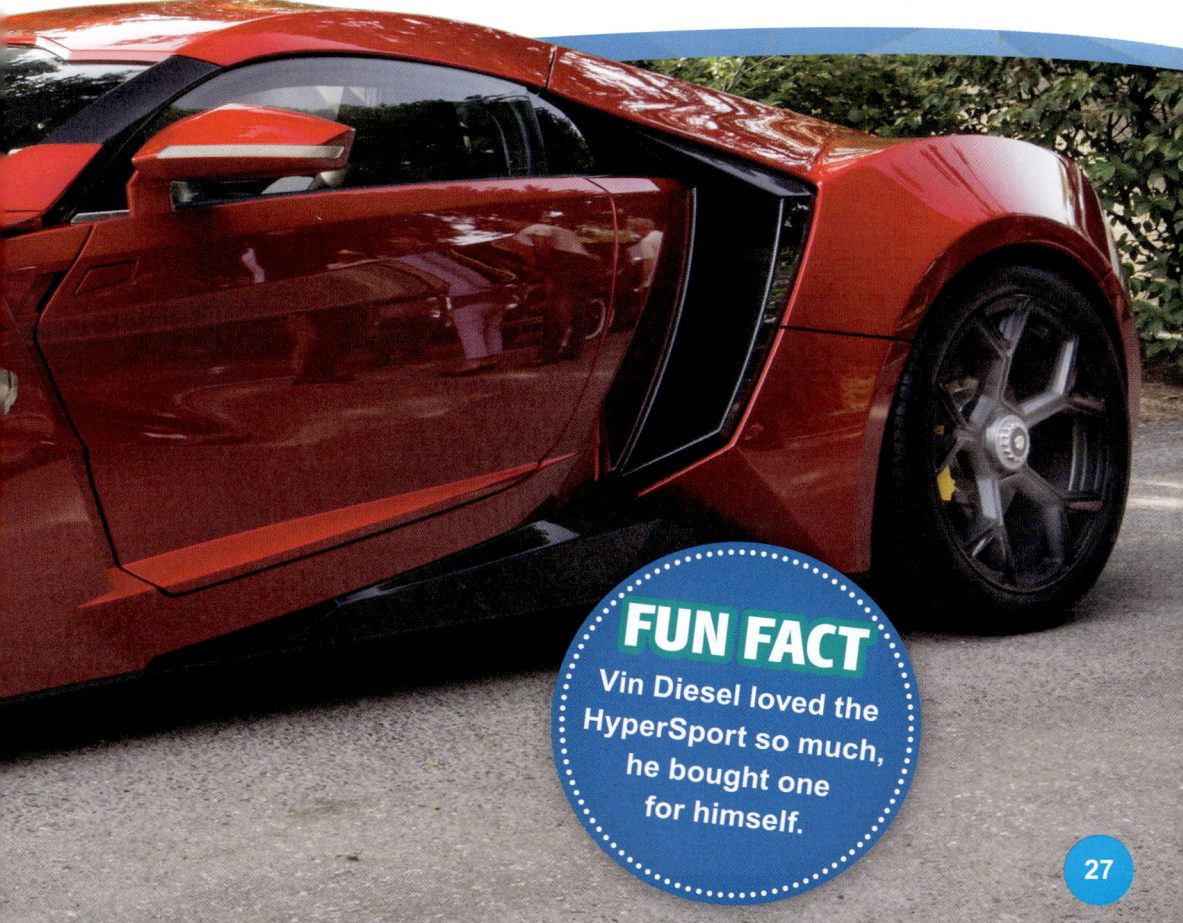

FUN FACT
Vin Diesel loved the HyperSport so much, he bought one for himself.

BEYOND THE BOOK

After reading the book, it's time to think about what you learned. Try the following exercises to jump-start your ideas.

RESEARCH

FIND OUT MORE. Where would you go to find out more about your favorite cars? Find out what company makes the car and locate its website. What information do the companies provide? What other sources of car information can you find?

CREATE

GET ARTISTIC. Cars start with creative artists and designers. Time for you to take a shot! Get art materials and create a great, new car. Will you make it a sports car? A sedan? A race car? What colors will you paint it? What features can you give it? Let your imagination go for a spin!

DISCOVER

DIG DEEPER. *Furious 7* made the Lykan HyperSport famous. For fun, pick your favorite movie and design a car the heroes could use. What features would it have? What would it look like?

GROW

GO TO A CAR SHOW. Car shows are a great way to see lots of cool cars up-close. Check your local events calendar, or ask at a car dealer for upcoming events. You can find shows of old cars and new cars, sports cars and classic cars. Go to a show and find a new favorite car to love!

RESEARCH NINJA

Visit www.ninjaresearcher.com/6112 to learn how to take your research skills and book report writing to the next level!

RESEARCH

DIGITAL LITERACY TOOLS

SEARCH LIKE A PRO
Learn about how to use search engines to find useful websites.

FACT OR FAKE?
Discover how you can tell a trusted website from an untrustworthy resource.

TEXT DETECTIVE
Explore how to zero in on the information you need most.

SHOW YOUR WORK
Research responsibly—learn how to cite sources.

WRITE

GET TO THE POINT
Learn how to express your main ideas.

PLAN OF ATTACK
Learn prewriting exercises and create an outline.

DOWNLOADABLE REPORT FORMS

Further Resources

BOOKS

Gish, Ashley. *Sports Cars*. Mankato, Minnesota: Creative Education, 2021.

Goldsworthy, Steve. *Scorching Supercars*. North Mankato, Minnesota: Capstone Press, 2015.

Storm, Marysa. *Supercars*. Mankato, Minnesota: Black Rabbit Books, 2020.

WEBSITES

Factsurfer.com gives you a safe, fun way to find more information.

1. Go to www.factsurfer.com.
2. Enter "Lykan HyperSport" into the search box and click 🔍
3. Select your book cover to see a list of related websites.

Glossary

aerodynamic: an aerodynamic design reduces the drag, or pull, on a car as it moves through the air. The Lykan HyperSport's spoiler is part of its aerodynamic design.

carbon fiber: carbon fiber is a very strong, lightweight material. Using carbon fiber to build a car makes it lighter and faster.

concept: a concept is an idea. A concept car shows off new features or designs.

hologram: a hologram is a three-dimensional image created by lights. The Lykan HyperSport's display panel is a hologram.

horsepower: horsepower measures the power of the engine. The Lykan HyperSport has 750 horsepower.

titanium: titanium is a hard, silver-colored metal. Titanium is stronger and lighter than steel.

transmission: the transmission is the part of the car that moves power from the engine to the wheels. The Lykan HyperSport has a six-speed transmission.

turbocharged: a turbocharged engine has added power. Turbocharged cars can go very fast.

Index

aerodynamic, 6
engine, 6, 19, 23, 27
Furious, 7, 8, 24, 27
gold, 19, 20
hologram, 19
horsepower, 6, 23
transmission, 23
turbocharged, 6
V6, 6, 23
W Motors, 12, 13, 14, 15, 17, 24

PHOTO CREDITS

The images in this book are reproduced through: Travel S/Shutterstock 21, Steve Lagreca/Shutterstock 26-27; All other images courtesy of W Motors.
Cover: Courtesy of W Motors, Mia Stendal/Shutterstock (background).

About the Author

Joanne Mattern has written many nonfiction books for children. Her favorite topics include sports, biographies, animals, and history. Joanne lives in New York State with her family and loves to drive fast cars.